The Man Who Created Paradise

The Man Who Created Paradise

A FABLE

GENE LOGSDON

Photographs by Gregory Spaid

Foreword by Wendell Berry

OHIO UNIVERSITY PRESS   ATHENS

Ohio University Press, Athens, Ohio 45701
Text previously printed in a limited edition, Cleveland, 1998
First Ohio University Press edition 2001
Printed in the United States of America

Ohio University Press books are printed on acid-free paper ⊗ ™

Cover photograph by Gregory Spaid. Cover design by Beth Pratt.

First paperback edition published 2017
ISBN 978-0-8214-2306-6

Library of Congress Cataloging-in-Publication Data

Logsdon, Gene.
    The man who created Paradise : a fable / Gene Logsdon ; photographs by
Gregory Spaid ; foreword by Wendell Berry. —1st Ohio University Press ed.
    p. cm.
ISBN 0-8214-1407-0 (acid-free paper)
    1. Restoration ecology—Fiction. 2. Farm life—Fiction. 3. Farmers—
Fiction. 4. Ohio—Fiction. I. Title.

PS3562.O453 M36 2001
813'.54—dc21

*The Ohio Art Council helped fund publication of this book with state tax dollars to encourage
economic growth, educational excellence, and cultural enrichment for all Ohioans.*

For
Wallace Aiken
&
John Gallman

Foreword

Maybe we continue to need to think of Paradise, and of making Paradise, because the earth as it was given to us (as we realize from time to time) was so nearly paradisal, and we are so talented at making a Hell of it.

Surely strip mining is the definitive sin of the industrial age. At least it is (so far) our most direct and deliberate act of Hell-making. We come to the coal-bearing slopes, rich on the surface with fertile soils and with forests. We find those soils and that forest—and all else we mean by "place"—to be in the way between us and what we want, i.e. coal, i.e. money. We therefore employ technologies more violent that earthquakes and avalanches to remove what is in the way, no matter that we destroy a greater wealth than we gain, and ruin a renewable resource for the sake of an exhaustible one. And then we foster and raise up the worst

Hell of all: a mind almost inconceivably narrow, which can justify this Hell-making as a necessity, a feat of economic progress, and a human good.

On the contrary, surely there is something wondrous and redemptive about a mind that can confront this definitive work of Hell of Earth Enterprises, Inc., and imagine the opposite story: How a member of the same species, out of his own horror at what has been done and his merely personal refusal to accept Hell as an acceptable human product, might employ the technology of destruction to begin the restoration of what has been destroyed; and how this singular effort might inspire the efforts of others to do the same thing; and how finally a whole community of people might ally themselves with the inherent goodwill of any place to heal itself and become the Paradise it once was.

This, then is a book of two visions: one of disease, one of health. Or to put it another way, Gene Logsdon has had the generosity and the courage to allow a vision of Hell to call forth in himself its natural opposite. But can we properly dignify the story of Wally Spero by the term "vision," or is it merely a reactionary fantasy? In my opinion, if you think this is merely a fantasy, you had better be careful. If you can look at the landscapes produced by strip mining without reacting toward some vision of the land restored, then you not only are looking at one of the versions of Hell; you are in it.

But can somebody really or "realistically" hope to accomplish what is accomplished in this story? Well, so far as I know, we don't yet have an example of a whole new community sprouting from the spoil banks of a strip mine. But it *is* possible for one inspired man and an old bulldozer to make a creditable beginning, as Gene Logsdon knows, because he has seen it, as I have myself.

Wendell Berry

The Man Who Created Paradise

*T*he letter stood out in sharp contrast to the others that fluttered across my desk regularly at *Farmer's Journal* magazine. Handwritten on yellow, lined tablet paper, it managed to convey in just a few words both fervent dedication and humor—a rare combination. The script slanted forcefully to the right in large, generous, yet angular, almost bayonet-like letters. I imagined the writer marching forward buoyantly but resolutely toward whatever life offered—the kind of personality one might expect from a man whose last name translated from Latin meant "I hope."

May 22, 1965

Dear associate editor Gene Blair,

Your article about how hybrid poplar tree cuttings will root and grow even on strip-mined spoil banks is exactly right. Isn't that amazing? I mean the poplar trees, not that you are exactly right. Know any other plants that would grow well on spoil banks?

I make farms. Alice helps a lot. Alice is my bulldozer. You should stop by and take a look at what we've done.

Yours truly,
Wally Spero
Paradise Road
Route 4
Old Salem, Ohio

I was used to getting letters from rural people who did not bother to give me enough details to grasp their situation clearly. In their intimate worlds, farmers knew the neighborhood details, no need to elaborate. And by habit, they tended to see everyone as a neighbor—able to "stop by sometime" even though I worked in Philadelphia, at least four hundred miles away from Old Salem, Ohio. But about the statement: "I make farms," I was mystified. If Mr. Spero was interested in spoil bank reclamation, I figured he must be using the bulldozer to level the banks or at least rearrange them into a more amenable landscape. But make farms on the strip-mined desolation of Appalachia? I had seen some of that land. One could sooner farm on the moon. I decided I would "stop by sometime."

Working as a journalist, even on a farm magazine, or perhaps especially on a farm magazine, had not given me much cause for hopefulness about what humans were doing to the planet. My work invoked in me only an angry sadness as I watched wealth and power, in the guise of "feeding the world," make land ownership, the lifeblood of democracy, more and more difficult for middle class people and impossible for poorer people. What coal companies did to Appalachia seemed to me no different than what agribusiness was doing to

farmland, only the coal companies did it faster—in years rather than centuries.

I should never have taken a job as an agricultural journalist in the first place. Interviewing people, let alone industrial farmers intent on getting rich through land expansion, and who therefore wittingly or unwittingly were puppets in the destruction of democratic society, was a trying experience for me. I was by nature a solitary person not inclined to minding other people's affairs. And the astounding energy of these large-scale farmers in pursuit of financial success both bored me and made me feel inferior.

But I hated even more sitting behind a desk editing vapid stories about money farming. As often as I could persuade my boss, I would travel farm country in the guise of an agribusiness reporter but hoping I would get sidetracked into something a little more inspiring. Knowing that I would hardly be allowed to spend any major time or expense writing about a man and his bulldozer on the agriculturally worthless spoil banks of Ohio, I used as an excuse for going to that region a dairy farmer who milked only twenty-five cows but earned an excellent income from them. My plan was to fly to Cincinnati, drive east and then north through Ohio's coal-stripped Appalachia, interview the dairyman near Barnesville, then head on north of Cadiz to find Wally Spero and fly back

from Pittsburgh at the end of the week. If I drove slowly, I could actually spend most of the time on the road between Cincinnati and Pittsburgh, contemplating my feelings of hopelessness and futility and trying to figure out what I ought to be doing with my life. All I really cared about was writing poetry and working a little farm, probably the two most unprofitable careers in America and so, for a poor man like me, impossible.

But far from alleviating my hopelessness, the ride became a travelog of despair. Strip-mined land was a biological horror, a farmer's nightmare more desolate than any bomb-pitted, warred-out battlefield. It seemed to me that no machines, however powerful and gigantic, could have reduced forested mountainsides to such an extensive moonscape of bare rocks and gutted ravines. Not even the fertile, narrow little valleys between the torn hills were spared, being dotted with ugly piles of cindery-looking stuff I would learn was called "red dog" on which nothing grew, and jagged jumbles of shale which supported only stunted weeds and brush. Crossing a bridge, I noticed that the water flowing under it appeared to be orange. I backed up and took a second look. The water indeed was orange.

"H'its from iron and sulfur in the water seeping out of old mine shafts," the serviceman at a gas station told

me. After forty more miles, I grew accustomed to seeing orange ribbons of water snaking through the green brush. Kind of pretty in a horrifying sort of way.

Where coal was evidently not close enough to the surface to be stripped out, and so the land left intact, loggers had invaded the mountainsides, leaving behind clearcuts that erosion turned into tumbles of huge boulders and a few frail saplings struggling to maintain a toehold. When rain fell, what little soil remained washed downhill, and the orange creeks turned pale brown temporarily.

At the foot of these raped hills, on the narrow strips of level land between the rutted roads and the orange creeks, what I took to be the third generation of once proud mountaineers—hillbillies—stood beside their house trailers that shook at the passing of every coal and lumber truck. They stared forlornly out at me from pinched faces pocked with soulless holes where happy eyes should have been. The men were sallow-skinned and generally skinny, with bulging neck veins and sharply protruding Adam's apples, nervous as rabbits in hunting season. The women on the other hand were mostly overweight, dumpy, hair long and stringy, with runty children hanging to them like baby opossums to their mothers. Invariably three rusting automobiles were parked beside

each mobile home, two of them jacked up on logs. Junk cars, worn-out tires, and beer cans littered the roadside between the residences and spilled over the creek banks into the orange water. I learned that people stared out at passing cars so pathetically because they saw the traffic as a symbol of escape—every car speeding down the road was a life raft away from their sinking ships.

Occasionally the winding roads led me through the ghosts of villages, rows of tar paper and tin, tall-legged shacks propped against the hillsides like old men hunkered against a barn wall at a farm auction, and a few modest but neat clapboard bungalows. "They git a real good welfare check," one man explained to me, nodding at a nicer house. The villages usually contained a dingy grocery store with a swinging screen door, the screen invariably wrenched loose from its framing above the door knob by countless hands pushing against it. Flies went in and out as unimpeded as people. There might be a gas station down the street, usually a church at the edge of town, and always a bar somewhere between. The church was most often a Quonset-type building, originally intended as a cheap substitute for a traditional barn. One evening I sat in the back of one such church and watched in near terror while people ran up and down the sawdust aisles in hysterical abandon, hurling themselves to the floor, loudly exhorting others in the congregation to escape the devil as they were doing.

Those who found no comfort in spiritual intoxication sought physical drunkenness in the saloon. In one of them I heard a man tell the waitress: "You gave me the clap, damn you." He was not angry but only expressing a fact—hopeless beyond anger.

The bar talk dwelt mainly on local robberies and fights, and on who had been laid off or hired at a shoe factory some twenty miles away, evidently the only chance for a job any of them considered. Someone's welfare check had been stolen and another's apparently lost in the mail. Mislaid or waylaid welfare checks were discussed in the same distraught tones that farmers in the corn belt used in recounting a foreclosure in the neighborhood. I climbed back into my rented car and kept on going.

The dairy farm I wanted to write about did offer some relief from the dreariness. The farm family was indeed making a good living from its hilly but well-kept little farm, giving a wonderful example of how Appalachia could be synonymous with prosperity, not poverty. But as quickly as my depression was relieved, as quickly it flooded back in response to what I learned. The monster power shovels were coming here too, to tear out the coal, said the dairyman, and there were fearful stories about people who refused to sell coal rights. "The strippers pushed dirt and rocks up against one man's property line until a heavy rain avalanched the spoil bank right down into his yard," the farmer said.

He meant to hold out, but what would it be like to have his farm surrounded by an alien moonscape?

I drove on with heavy heart, trying now to avert my eyes when passing coal-gutted regions. That was difficult, almost impossible, as I wended my way through the country around Cadiz. I tried to concentrate on the radio to avert my attention from the landscape. Little help there either, as one after another plaintive sound of "country music" deplored unrelenting poverty and fleeting, comfortless sex.

> *Sixteen tons and what do you get? Another day older and deeper in debt.*
>
> *If you've got the money, honey, I've got the time.*

Old Salem's "business district" consisted of a general store that served also as a post office, a welfare office, and five empty, boarded up buildings. When I mentioned Wally Spero's name in the store, a gleam of more than recognition shone in the eyes of the loungers by the wood stove. Was it pride?

"Wally? Well, yeah, I know where he lives. He is quite a Wally now, I can tell you that. You just take the road up yender to the north. You'll come to what was once a schoolhouse, couple miles up, and an abandoned church across the road. All falling down. Right beyond, there's a gravel road to the left with a hand-painted sign that says Paradise Road. You just follow that till it quits and then you're there."

"There?" I asked, puzzled.

"Yep, then you're in Paradise, and Wally will be around somewheres. If you can't find him, listen for Alice. You know Alice?"

I smiled and nodded, thanked the men and proceeded on my way.

I had not written ahead to warn Mr. Spero I was coming, like a journalist ought to do. People I hoped to write about I wanted to happen upon as if by chance, if not actually by chance. I wanted to talk to them like a stranger on the way to becoming a friend, enjoying the moment, no thought of any ulterior journalistic purpose in either person's mind.

Going north, "up yender," I again was overwhelmed by strip-mined country on both sides of the road. The spoil banks here measured about sixteen feet high and twenty feet wide at the bottom, parallel to each other in

a regular, distinct corduroy pattern, like giant windrows of hay on a hillside field. It seemed as if an earthquake had shaken these hills until their earthen skin rippled like water and then froze solid into wrinkles that would remain until the next glacier. The size of the overgrowing thorny black locust trees dated the spoil banks at about twenty years old. On occasional ledges that had been left undisturbed between the banks, trees of moderate size and even a few old gnarled leftovers from a century ago held their ground.

A pig chewing on acorns stood incongruously in the doorway of the dilapidated church. It looked as satisfied with its acorns as a minister with a good Sunday morning collection. There seemed to be no farmstead around to which the animal belonged. No doubt it, or its forbears, had been left behind when the last farmer pulled up stakes, and had gone wild. Across the road, a groundhog stuck its head out of a hole in the foundation of the crumbling schoolhouse and watched me curiously.

I had no trouble seeing the sign marking Paradise Road a little farther on. The letters stood out in yellow against a blue background, as bright and sassy as an early spring windflower blooming beside a snowbank. It marked the entrance into a barely discernible lane that had clearly been gouged through the spoil banks up the hill with a bulldozer. After going about a thousand feet up the lane, which quartered the soil banks like a boat in heavy seas, I drove out on the top of a forested ridge that had been left intact by the strippers. What then confronted my eyes caused me to cry out in surprise and pleasure.

In the valley below, plunked down into this grim landscape of spoil banks, lay as pretty a little farm as the most imaginative Currier & Ives artist could have conjured up on canvas. For an area that I guessed covered about fifty acres, the parallel ripples of spoil bank were interrupted and replaced by little fields of verdant grass quilting the hillsides going down to the valley floor and again up the hill on the other side of the valley. Woven wire fences divided the fields. Small hillside ponds, filled by runoff water from the pastures, and often surrounded by little groves of trees, adorned the meadows. A creek threaded the valley and the water looked blue, not orange. The narrow bottomlands on both sides of the creek were laid out in little fields too, but here annual crops were growing, corn and oats and clover almost as lush as what I had noticed at the dairy farm I had visited. Where hillside met bottomland, a stone and wood barn had been built into the hill, so that there was a drive-in entrance directly to the second floor on the uphill side

of the structure, and on the valley side a drive-in entrance to the ground floor. About five hundred feet from the barn and a bit higher in elevation stood a little stone-walled house tucked into the hill in the same way. The roof was covered with hand-cloven, red oak shingles. A white picket fence surrounded the house and gardens, and dark plank fences bordered the barn lots. Flowers bloomed everywhere. Cattle and sheep grazed in the hillside pastures. Chickens cackled in the barnyard. The big letters on the barn spelled PARADISE FARM and to a countryman's eye, this scene did indeed look like paradise.

After I had paused long enough to realize that I wasn't looking at a mirage of my own deep desire, I could see, as well as hear, a bulldozer clanking down out of the spoil banks next to the highest field across the valley. The person on the bulldozer had spotted me, and was waving and motioning down toward the barn where he was apparently headed. Although there was a lane of sorts to the house, I decided to walk down, so as to take in the details of the farm better.

The trees in the pasture were mostly oak, thornless honey locust, hickory, wild cherry, sugar maple, and black walnut, all providing not only shade but food for man or animal and all producing valuable wood. Passing

one of the hillside ponds, I heard bullfrogs and saw a school of fish lurking at the shoreline. At another, mallard ducks, half-tame, floated serenely on the surface.

Everywhere were signs of meticulous and calculated work. The fence posts were black locust and cedar heartwood, native to this country and capable of outlasting steel ones. The end posts, of the same two woods, were squared, twelve inches thick, each with a brace post nearly as stout. They would still stand solid forty years from now. The fence was stretched so tightly that the wires hummed in the mountain wind. A top strand of barbed wire, nailed to the posts with staples, ran four inches above the fence and almost perfectly parallel to it.

The fence lines marked changes in the slope gradients as accurately as elevation lines on a topographical map. The upmost fence separated the forested ridge top from the steep-sloped permanent pastures of bluegrass, lespedeza, and white clover on the middle slope of the long hillside. The next fence, downhill and roughly parallel to the first, separated the permanent pasture from the temporary pasture and hayfields on the bottom slopes, which were less steep and so able to endure an infrequent cultivation for new seedings of oats, red clover, alfalfa, timothy, and orchard grass. The third parallel fence line

at the foot of the slope separated the temporary pasture fields from the level narrow valley fields that could be cultivated to grains and clovers annually without net soil loss. Cross fencing, running longitudinally down the hillside, further divided each of the three latitudinal sections into smaller fields yet, and it was apparent that Mr. Spero not only rotated his annual crops from year to year, but rotated his livestock frequently from one pasture to another too, so that the grazed plants were kept in vigorous, palatable condition with the least amount of mowing but without overgrazing. I could not see one thistle, sour dock, or patch of poverty grass anywhere on the grassy meadows.

By the time I reached the barn, Mr. Spero was waiting for me, still seated on his bulldozer, a huge old Allis Chalmers HD 19. He was smiling broadly, his usual expression, I soon learned, even though I had come to his place unannounced.

"I'm Gene Blair from the *Farmer's Journal,*" I said as quickly as he had turned off the bulldozer's rumbling engine. I felt, as usual, both tense and embarrassed even though I had an invitation of sorts to be here.

"Well, I'll be switched," he said. "I never thought you'd really come." He spoke in a high-pitched voice, evidently the result of being habitually excited. Peeling

out of the bulldozer seat, he strode over to me, sticking out a callused, stubby hand of welcome.

Because I was still under the spell of astonishment that this strange over-the-rainbow farm evoked in me, I could not contain myself through the conventional small talk of first meetings, but straightaway blurted out: "I'm not sure this is all real. Was all this spoil banks?"

He grinned like a schoolboy who has gotten away with putting a dead mouse in the teacher's desk drawer.

"It's the most funnest thing I ever did do," he said. "I bought the land for about five dollars an acre. Even on foot you could not get through it and the owner thought I was nuts. I've been playing Rembrandt for seven years. Alice is my artist's brush and the spoil banks are my canvas, and I just paint fields on it."

I could think of nothing to say, so he continued. "Yep. Started when I was twenty-four. I try to make a quarter acre of farm every day. Course it don't always go that fast. Have to paint in ponds and tree groves at suitable places and cut out salvageable trees for lumber and posts and fuelwood, and replant more trees, and gather up rocks and build the house and barn and then start farmin' when the buildings were finished and the first fields grassed. And some days Alice gets sick and needs to be operated on. She was near dead when I found her on a dealer's back lot and I expect I got five thousand dollars in repairs in her over seven years. But no way else could I afford to own a bulldozer. Alice doesn't mind livin' low on the hog. In fact we took our last savings and bought a couple more hundred acres to make more farmland with some day. It's the only way a poor man can own a farm that I know of. Make it yourself. Some days I think I'm God."

He proceeded to show me around his paradise and the wonders he had created. There were boulders in the barn foundation that five men couldn't lift, but which Alice had pushed and nudged gently into place. "Not even an atomic bomb could dislodge that som'bitch," he said, nodding at his barn with great satisfaction. "All the stone for the walls was right here. I just had to develop an eye for which rock oughtta go where. Then fill the cracks between with concrete to lock 'em in place. It was the same with the timbers. There was a lot of young black locust on the spoil banks—locust is a legume, you know, and can grow about anywhere because it provides its own nitrogen from the air. Anyway black locust doesn't rot much in the ground and hardly at all above ground. So before I'd level a section of bank, Alice would shove the brush aside and I'd cut down all the locust that was eight inches in diameter or

more. Split the smaller ones for posts and squared the larger ones for timbers."

As we walked from field to field, from one farm animal to another, almost from one tree to the next, it was apparent that Wally Spero had at his fingertips a remarkable fund of knowledge about traditional farming and gardening. Had he grown up on a farm?

"No. I can't even tell you how I got interested. I was working as a metal grinder in a foundry—dirty work but good money. I just couldn't see bowing and scraping to a boss all my life and being totally dependent on that job, no matter how good the earnings. I started reading about farming, first just by accident. I realized right away that there was a possibility of making enough money to live independently on a farm, once the land was paid for. That possibility seemed like heaven on earth to me. I searched out books and magazines, everything on subsistence living. Even at work on my breaks, I'd read. The guys made fun of me. I asked them if they ever got worried about their food supply, or ever had any notion that their lifestyle might be in danger. They would just stare. I told them someday I'd walk out of that place a free man. They laughed. But I started singing inside my grinder's mask. I had figured out my escape route. I found out that strip-mined land could be bought real cheap. I looked at this stretch of it and realized there were possibilities no matter how forbidding the land looked. The stripping had not gone so terribly deep here and there was topsoil buried in the banks. I stayed on at the foundry the first two years that Alice and I started making a farm. We did a little every day we could, weekends, holidays included. And then one fine spring morning when I needed to seed my first fields, I walked out of the foundry, just as I had planned, and I never went back. I remember how curiously the fellows stared at me on that last day. They couldn't believe I'd really sprung the trap. They didn't think that was possible."

We talked all night, and I left early next morning to catch a flight from Pittsburgh back to Philadelphia. I was so excited that although I was exhausted, I could not sleep on the plane as I usually did. Even the reality of taking a cab from the airport into the smoking, bustling, crowded city did not check the enthusiasm for another life that Wally had fired up in me. I thought of him singing inside his grinder's helmet and I started humming inside the cab.

Next day, I wrote a passionate story about Wally Spero and Alice. It gathered dust on the Managing Editor's desk a week before he brought it back to me, and hesitantly, trying not to hurt my feelings, said the

story did not "compete" with the news that mattered to farmers. I threw it in the wastebasket and stared out the window the rest of the day.

Wally and I corresponded regularly for about a year although I could tell that letter-writing was not something he enjoyed. He was continuing to "make farms," he wrote, and said a few other people were talking to him about trying to do it themselves. The last time he wrote was to tell me that he had met a woman willing to share his way of life—something he had worried might not occur—and they were soon to be married.

In the meantime, the news that "mattered to farmers" evidently didn't matter enough, because the Farmer's Journal Publishing Company began to experience a slow but clearly discernible decline in revenue that forced a parallel slow but steady attrition of staff. Eventually my position, precarious at best, was extinguished without formal warning, although I was not surprised. Somewhat relieved, I went to work for a magazine called *Ecological Order* and leaped into the environmental fray, thinking, erroneously as it turned out, that all those words I now began to milk vigorously from my typewriter would do some good. I lost track of Wally and came to think of his little farm as an aberration, the result of an errant gene in one person and no other, a gesture as vain before the dreadnought of environmental destruction as my writing was proving to be.

The years slipped by as they must do—oh, so terribly fast. In 1994, weary of fighting the environmental battle with words, I decided to retire to my own rural retreat and as much as my advancing age would allow, to do battle by action, fashioning my own little farm out of some cheap, rundown cornfields in midwestern Ohio. I did not need a bulldozer; just clover and patience. I began to think again of Wally Spero who had engendered this idea in me and of what might have happened to his farm. My common sense suspected that I would find it now, thirty years later, abandoned to multiflora rose. Or sold for a landfill or an incinerator or a so-called low-level nuclear waste disposal site, since Ohio's Appalachia more and more had become a dumping ground for the society that had raped it of its riches. No reply came from a letter sent to his postal box address. I could dig up no telephone number for him. My fears grew.

Finally curiosity overcame me and, finding myself on a trip through eastern Ohio, I detoured to Old Salem. Nothing I saw along the way suggested anything had changed for the better. Appalachia was still Appalachia, and although in some areas, especially along highways, the coal companies had done some remarkable

reclamation, no pattern of thrifty family farms had re-appeared on this land. And the villages in the hollows off the Interstates were deader than they had been thirty years ago. The number of junked cars, old tires, and beer cans had increased. And there were still creeks running orange water. What was new was a horizon of polluted air: coal-burning generators; nuclear cooling towers; a huge incinerator smokestack; oil and gas refinery smog. The wasting of Appalachia had proceeded now until it affected even the air above it.

So I was shocked when I drove into Old Salem. It had a new and vibrant look about it. The general store had been spruced up almost beyond recognition and four previously boarded-up storefronts had been re-furbished for businesses. One was a pizza parlor with a sign that read: "Homemade bread every Saturday / Homemade ice cream every Wednesday." The sign above another store announced: Food Fresh From Farmers: meat, milk, cream, butter, eggs, vegetables, fruit, and grains. A third store sold "Mountains of Good Used Clothing" and "Old Salem's Own All-Purpose, Low Cost Moccasins. Made Right Here. $20 a pair." The fourth business, Small Farm Supply and Repair Co., sold seed and seeders, tillers and mowers, pitchforks and spades, sheep shearers and milk buckets, chicken waterers and axes, all manner of tools and aids in raising crops and animals on a small scale. The business had even added on a second, new building in which smaller old tractors and tillage tools, horse-drawn equipment, and, strangely enough, a dozen or more old bulldozers were being repaired and restored. The last building at the end of the street, a Quonset-type structure, was newly painted and spotted a large sign that read: "Auction Every Saturday." In the adjacent lot, lined up and apparently awaiting the next auction, was an assortment of wooden troughs, hay racks, sheep mangers, gates, fence posts, tiny chicken coops, sow huts, and large, long-tined wooden scoops that obviously fit on tractor front-end loaders to pick up hay out of windrows—a variation on the buck rakes of my youth. All were obviously homemade from local wood. A man putting tag numbers on the various items laughed when I asked him if some sort of traditional old timer's festival was in progress. "Nope," he said. "But a fancy lady from Pittsburgh stopped in last week and bought one of those little sheep pen hay racks. Said it would make a 'daaahling' magazine rack in her new home."

I could barely resist the temptation to spend the whole day in the town, not only because I was overcome with curiosity as to what had happened but because these stores in one way or another were serving exactly the kind of lifestyle I was trying to live. But stopping proved nearly impossible because every second of forward travel brought new surprises luring me on. The road, for one thing, had been newly black-topped. On each side of the highway, a string of modest but neatly kept houses had been built. Most startling of all, the once tumble-down church had been restored and a spire added to the roof. A sign out front said "Welcome, If You Believe In Paradise." The little school across the road had been completely rebuilt, and children were playing in the schoolyard. What in heaven's name had happened here?

The blue and yellow Paradise Road sign was still there, though I was sure the original one had long since been replaced. The road going up the hillside had been widened and heavily graveled. The spoil banks along both sides had been roughly leveled, and a dense stand of evergreens hid the scars. I drove up to the ridge with heart pounding, expecting the best, expecting the worst.

But no hope of great expectation could have pre-pared me for what came into view. Wally Spero's place was indeed still there, just as it had looked thirty years ago only the trees in the groves were of much larger girth. But what took away my breath now was a whole little world of Spero-like farms spread up and down the valley

as far as eye could see! I thought of a suburban development, only instead of just big, expensive, hard-to-heat, look-alike houses flank to flank, this "development" consisted of scores of mini-farms of varying sizes, each with a small, energy-efficient farmhouse, barn, chicken coop and other outbuildings, fields, gardens, and ponds. On many of the houses were solar electric generating panels. And unlike the usual suburb, which generally appears deserted most of the daytime, there were adults and children scattered all over these tracts, all busily at work or play.

Narrow gravel roads connected the farms, and I could drive them if I went slowly, although it occurred to me that there were no other cars in evidence. I drove into Wally's barnyard to find a lean and wiry man there with his head stuck under the hood of a bulldozer that looked like Alice. The man turned, and although he was bald now, I could recognize Wally from his wide, toothy grin. Alice had changed not at all.

If I had been surprised and elated at our first meeting, I was now beside myself with such astonishment that I could hardly speak coherently. After but a bit of hesitation, for I had grown gray and sag-jowled, Wally remembered me, and began to talk in his characteristic nonstop enthusiastic way.

He had gone on making farms whenever his own did not otherwise need his time and energy. "I can't really explain what happened," he said, still the mischievous schoolboy grin on his face. "People just started showing up. They'd stand around and stare and go away. And come back again. One day, a fella offered to buy some land that I had about finished 'painting' into a farm, and not so long after, another guy wanted to buy a piece though I had barely started leveling the spoil banks on it. He had his own bulldozer. Then things just went bonkers. Exponential growth. I had attracted the first two, you see, and they each attracted two or three more, and they in turn each attracted several more and we all kept on attracting still others and pretty soon there were people and old bulldozers crawlin' over these hills like a bunch of tumblebugs on a giant cowpie. I declare it's been the most funnest thing I ever did do. One strappin' young fella who had given up a promising career as a baseball pitcher didn't have a bulldozer, just a team of horses and a slip-scraper. He painted himself a seven-acre strawberry farm and he's been makin' enough from it for all his cash requirements. I'm tellin' you, people aren't dumb or lazy. They just gotta see the possibilities—understand that they can do it. Then get outta their way. Cmon, I'll hitch up a horse and show you around."

Horse? This was too much for me. Why would a man who lived by bulldozers keep a driving horse? I sputtered as much out loud. "It's more logical than you think," Wally answered. "First off, not many folks here hanker to travel much beyond the next ridge and I definitely don't. We've got everything we want right here. A horse will get you to Old Salem almost as quick as a car, and a lot quicker in winter and spring when our little roads are hard to negotiate. And this way we don't have to spend zillions of dollars to build roads. Four-wheel-drive pickups would be nice but who can afford 'em? Some people drive to town on their tractors." He laughed heartily, obviously seeing great humor in that. "Well, why not?" he challenged me.

And I had to agree. Why not indeed.

Wally's driving horse turned out to be a Belgian draft horse, but no matter. It clip-clopped along just fast enough to see this country the way it needed to be seen. Wally took me from farm to farm, talking all the while, introducing me to everyone we met. That meant slow going, since someone was at home at nearly every place we passed, working in their fields or barns or busy in shops or offices. All had time to stop and talk. I could scarcely believe the variety of work in progress. Some worked in home offices for businesses far away or were putting out catalogs that offered their neighbors' home-produced goods nationwide. Many were crafting a wide variety of wooden materials, mostly furniture but also toys, gun stocks, woodenware, fencing, archery bows, and boats. One man had found a small but steady market for persimmon wood golf club heads. "I net seven thousand dollars a year from them, all I need along with the farm production," he said. Another was growing bamboo, cutting and curing it for a variety of uses, such as bean poles, garden stakes, electric fence posts, lawn and patio furniture, and even fishing poles.

"There are hundreds of little farm ponds in these hills now, most of them full of bass and perch that make as fine a dish as any restaurant serves," Wally said, laughing again at the great humor he saw in the situation. "It's about the only fish you can get that doesn't come out of polluted water. We've all got clients who pay good money to fish our ponds because of that reason or because it's the only way you can get really fresh fish. The ponds are mostly small so the most convenient way to fish 'em is with a simple bamboo pole like in the old days. Or with a great big seine if you want to sell a bunch at one time."

The mini-farms were teeming with cottage industry. Potters were at work firing up their kilns. A winemaker was making a living from a fifteen acre winery. Spinners

and weavers labored at their wheels and looms, turning the wool from their sheep into clothing and blankets. "Sheep are the perfect farm animal for us," Wally said. "They can be raised on grass and hay alone, without disturbing the land with annual grain crops and causing erosion. As they graze, sheep do the harvesting themselves and spread their manure too. They provide wool, the best fabric for clothing. And hard to beat a lamb chop for good eating. We are now learning how to breed sheep for home milk production too, and to sell for Roquefort cheese. Now that's what I call an all-around animal."

I saw not one place that did not have a small flock of hens. "We hold an auction every Saturday in Old Salem now," Wally said. "Everyone sells the surplus production from their farms there plus the other home-manufactured stuff. It's drawing a big crowd, both buyers and sellers. Even if you've only got one dozen eggs, or had time to bake only two loaves of bread, or picked one basket of wild elderberries, you can sell 'em at the auction. It's just the most funnest thing I ever did see." He then gave me a sly wink. "If you don't raise or make it yourself the auction won't handle it. That was my idea."

Remarkably, he never made grand statements or conclusions or generalities about Paradise. He did not

say that here was an example of New Age Economics; or that Paradise represented a rejection of both Socialistic and Capitalistic Totalitarianism; or that the homesteaders of Paradise were the New Pioneers escaping and eventually replacing the dying society of Mall People. Nor did he claim that Paradise was a return to basics, a return to roots, or a return to the simple life. "It ain't a return to anything and it ain't simple," he said. "It's going forward and it's very complex." He had no intellectual theories about what was going on here or why, nor did he seem to think any were necessary. He only made particular observations about particular people doing particular work. "See that fellow over there making a haystack?" He pointed out to me as we rode along. "See that big wooden fork on the front-end loader of his tractor? Now that's a story. Hay and pasture are the mainstays of a truly sustainable kind of farming. No erosion involved. Not cultivating annual grain crops at all, if possible. We can raise cattle and sheep on pasture alone, with surplus hay to tide us over winter. But since money is very tight here, we have to use as little of it as possible and so in this case it becomes a matter of what's the cheapest way to make hay on a small farm. Balers are expensive. Old Ned Kottering got to remembering how when he

was young his father scooped hay out of the windrow with a giant wooden fork fixed to the front of an old car or truck and then hauling the scoop-fulls to haystacks where a mechanical stacker lifted the hay to the necessary height. Well, he figured, modern hydraulic front-end loaders on tractors could be modified to do that, and one man with just a few acres could scoop and stack his hay crop without help and very cheaply. Now nearly everyone does it. The cows and sheep eat the hay right out of the stacks. Don't hardly need a barn anymore."

He chuckled and continued: "One guy did bring in one of those big round balers once. The fourth bale he made came out of the baler and shifted around somehow to face downhill. Before he could get to it, it started rolling down the mountain. Took out twenty rod of fence and a shed before it finally stopped in a pond." Wally thought that terribly funny and whooped uproariously.

He described another farm we were passing. "That woman's a real character. Genius really. She has an ever-flowing spring, and grows watercress for sale in the crick that flows from it. She also sells ginseng and goldenseal from the woods and raises snapping turtles in her pond. Ever eat fried snappin' turtle? Gawd, it's heavenly. A restaurant takes all she can't sell or trade locally."

All the while, even after Wally introduced me to his wife, their two married children who were raising families on their own homesteads in the valley, and the grandchildren, I felt that there was something perhaps even more wondrous that he was holding back from me. Finally, late in the afternoon, he turned the horse into a road we had not previously traveled. "I want to take you over the ridge to the next valley," he said, nodding westward, the mouse-in-the-desk grin appearing again. "This will pop your eyes out." The road meandered up to the top of the ridge that I thought marked the western border of Paradise. At the very summit, we came out onto a clearing among the trees that allowed us to see out across the next valley and on to the horizon of yet another ridge several miles away. Here, instead of a pattern of homesteads replacing spoil banks, were tracts of evergreen trees in varying stages of growth quilting the once-torn hillsides, with a few houses and gardens spotted here and there among the young groves, all radiating out from a very large building overlooking the valley from the opposite ridge. I could feel Wally's eyes on me, enjoying my astonishment.

"A very unusual bunch of people bought up this land—three thousand acres of it—and moved in here about fifteen years ago," he explained. "It ain't a commune

exactly. They all own their own places, but they got this Christmas tree farm business going as a cooperative venture. They call the place Raven Mountain. But the trees are only one of their businesses. You'd never in a million years guess what they make in that building." He didn't wait for me to venture a guess. "Compost toilets. Can you believe that. Waterless toilets. The things really work. Worldwide sales. Even folks here are getting them. Beats the hell out of a plain old privy. Ain't it somethin'. And now they are starting to make solar hot water heaters and all kinds of solar electric gimmicks. I tell you, this has been the most funnest thing that I ever did see happen."

As of this writing, Paradise is still expanding as fast as clattering old bulldozers can move, and Wally Spero's hand and spirit still enliven the mountains. Land speculators opened a couple of offices in Old Salem with the idea of developing some "unspoiled prime housing locations" in the area. But it soon became evident that no one in Paradise would sell and all the land around it was owned or controlled by something called "Alice Inc." which, so rumor has it, regularly turns down million-dollar offers for condominium and ski resort sites, preferring to sell small acreages at fifty dollars per acre to poor people willing to work hard—with a clause in every deed, Wally says with his dead mouse grin—that the land must be sold back to Alice Inc. if the homesteader decides to leave.

Recently, several government agencies enthroned themselves in Old Salem intending to administer handout programs to the poorer homesteaders and regulatory programs to the successful homesteaders. The bureaucrats, like the land speculators, eventually closed for lack of business. In fact, in one of its periodic cost-cutting moods, the Department of Health and Human Services closed the welfare office in Old Salem. Try as they might, social workers could not find enough people in need to justify keeping the office open.

Something even more ironically amusing has occurred, at least amusing to one bald-headed old man still leveling spoil bank dirt when neither Alice's engine nor his arthritis is acting up. The Universal Electric Power Co., which twenty years ago would not bring utility lines into Paradise except at a price none of the homesteaders could afford, now, with the increase in population, wants to supply the needs of the community. But no one will sign up. After the early years of doing without electricity, Paradise has equipped itself with solar panels, windmills, and diesel generators to provide all the electricity its people feel they need. The old bald-headed man on his bulldozer, with a sassy little granddaughter nestled in his lap, cackles. "It's just the most funnest thing that I ever did see."

Photographs by Gregory Spaid